I0471102

PREVENTION THROUGH DESIGN
PLAN FOR THE NATIONAL INITIATIVE

DEPARTMENT OF HEALTH AND HUMAN SERVICES
Centers for Disease Control and Prevention
National Institute for Occupational Safety and Health

DISCLAIMER

Mention of any company or product does not constitute endorsement by the National Institute for Occupational Safety and Health (NIOSH). In addition, citations to Web sites external to NIOSH do not constitute NIOSH endorsement of the sponsoring organizations or their programs or products. Furthermore, NIOSH is not responsible for the content of these Web sites. All Web addresses referenced in this document were accessible as of the publication date.

ORDERING INFORMATION

To receive documents or other information about occupational safety and health topics, contact NIOSH at

Telephone: **1–800–CDC–INFO** (1–800–232–4636)
TTY: 1–888–232–6348
E-mail: cdcinfo@cdc.gov

or visit the NIOSH Web site at **www.cdc.gov/niosh**.

For a monthly update on news at NIOSH, subscribe to *NIOSH eNews* by visiting **www.cdc.gov/niosh/eNews**.

DHHS (NIOSH) Publication No. 2011–121

November 2010

SAFER • HEALTHIER • PEOPLE™

FORWARD

Designing out hazards is the most effective means of preventing occupational injuries, illnesses, and fatalities. Although this concept is well known, there has not been a concerted effort to achieve broad implementation of it. In 2007, NIOSH initiated a national initiative, Prevention through Design (PtD), to foster designing out occupational hazards in equipment, structures, materials, and processes that affect workers. Drawing on the knowledge of many different stakeholders, NIOSH is now presenting the *Prevention Through Design Plan for the National Initiative*. This is a plan for the nation and will focus on specific goals and activities in areas of research, education, practice, policy, and small business. NIOSH will continue to seek partners to implement this plan. By having specific goals and activities, the national effort can be focused to more effectively protect workers from occupational injury and disease. Collectively, achieving these goals will influence the change in culture that is necessary to include PtD principles in all designs affecting workers.

John Howard, M.D.
Director, National Institute for
   Occupational Safety and Health
Centers for Disease Control and Prevention

PtD Council Members

Michael Behm
East Carolina University

Donald Bloswick
University of Utah

Theodore Braun
Liberty Mutual

Wayne Christensen
Christensen Consulting
  for Safety Excellence, Ltd.

Noah Connell
Occupational Safety and
  Health Administration (OSHA)

Lucy Crane
Whole Health Management

June Fisher
Training for Development of
  Innovative Control Technologies
  Project

John Gambatese
  Oregon State University

Charles Geraci
  NIOSH

Donna Heidel
  NIOSH, PtD National Initiative
  Coordinator

David Heidorn
American Society of Safety Engineers

James Howe
Safety Solutions

Walter Jones
Laborers' Health and Safety Fund
  of North America

Mary Ann Latko
American Industrial Hygiene Association

Mei-Li Lin
National Safety Council

Joanne Linhard
ORC WorldWide™

Bruce Main
design safety engineering, inc.

Fred Manuele
Hazards Limited

James McGlothlin
Regenstrief Center for Healthcare Engineering

John Mroszczyk
Northeast Consulting Engineers, Inc.

Margaret Quinn
University of Massachusetts Lowell

Danezza Quintero
OSHA

Andrea Okun
NIOSH

Michael Pankonin
Association of Equipment Manufacturers

James Platner
Center for Construction Research and
  Training (CPWR)

Scott Schneider
Laborers' Health and Safety Fund
  of North America

Paul Schulte
NIOSH

Richard Sesek
University of Utah

Erica Stewart
Kaiser Permanente

John Weaver
Birck Center for Nanotechnology Research

Frank White
ORC WorldWide™

Dee Woodhull
ORC WorldWide™

CONTENTS

Summary

In 2008, among U.S. workers, 5,071 died from occupational injuries, 3.7 million suffered serious injuries, and 187,400 became ill from work-related exposures [BLS 2008]. The estimated annual direct and indirect costs of occupational injury, disease, and death range from $128 billion to $155 billion [Schulte 2005]. While the underlying causes vary, a recent study implicates design in 37% of job-related fatalities [Driscoll et al. 2008]. Thus, to protect lives and livelihoods, stakeholders across all industrial sectors of the economy need a comprehensive approach for addressing worker health and safety issues by eliminating hazards and minimizing risks to workers throughout the life cycle of work premises, tools, equipment, machinery, substances, and work processes, including their construction, manufacture, use, maintenance, and ultimate disposal or re-use.

The following document provides the rationale, mission, objectives, outcomes, and timeframe for the Prevention through Design (PtD) National Initiative.

The mission of the PtD National Initiative is to prevent or reduce occupationally related injuries, illnesses, fatalities, and exposures by including prevention considerations in all designs that affect individuals in the occupational environment.

This will be accomplished through the application of hazard elimination and risk minimization methods in the design of work facilities, processes, equipment, tools, work methods, and work organization. Although the ultimate goal is to "design out" potential hazards at the beginning phases of a project, rather than dealing with problems inherent in completed systems, PtD methods also can effectively be applied to existing processes and equipment. Eliminating hazards and minimizing risks during the design, redesign, and retrofit of facilities, work processes, and equipment may ultimately save money and, more critically, will protect workers.

PtD is a national initiative; the nation must focus its collective attention on eliminating hazards and minimizing risks to workers and the work environment, because no single organization or occupational discipline can achieve its goals. Success will come through a coordinated, phased approach to PtD activities that takes into consideration the unique challenges faced by businesses within all industrial sectors. Through the collaborative efforts of industry, labor, professional organizations, academia, government agencies and PtD experts, and with the commitment of professions affected by PtD issues (including architects, industrial designers, and engineers; purchasing, finance, and human resource professionals; and health and safety experts), the PtD National Initiative can save lives and demonstrate financial value.

The PtD Council, comprised of a diverse group of individuals from industry, labor, academia, and government agencies and other PtD experts, drafted the goals, activities, and timeframes set forth in the PtD National Initiative. The Council based its work on input received from approximately 250 stakeholders at a PtD Workshop held on July 9–11, 2007. The initiative's goals are organized around four overarching areas: *Research; Education; Practice; and Policy. Small Business* was added as an additional focus for goal development to address the unique challenges of applying PtD methods to small business processes and environments. Each of these overarching areas, as well as the small business focus area, is supported by a strategic goal. A summary of the strategic goals for each of these areas is provided below. Details about specific activities for accomplishing the goals, as well as performance measures and timeframes, can be obtained by clicking on the specified links.

1. Research: *Research will establish the value of adopted PtD interventions, address existing design-related challenges, and suggest areas for future research.*

    Research focuses on design factors that effectively reduce occupational morbidity, mortality, and injury; metrics that assess the impact of these design factors; methods that diffuse effective designs; and economic and business issues, including financial analysis of the impact of Prevention through Design on the business process.

2. Education: *Designers, engineers, machinery and equipment manufacturers, health and safety (H&S) professionals, business leaders, and workers understand PtD methods and apply their knowledge and skills to the design of facilities, processes, equipment, tools, and organization of work.*

    Education focuses on motivating and equipping professional communities to continually increase their knowledge of design features that have a positive impact on worker safety and health. Acquisition of PtD knowledge and skills will occur through enhanced design and engineering curricula as well as through improved professional accreditation programs that value PtD issues and include them in their competency assessments. Making business leaders aware of the potential for increasing company profitability by incorporating PtD methods in their systems and processes is also an important component of the education goal. Further, some engineering educators are now advocating revision of curricula, so the time for incorporating PtD is potentially at hand.

3. Practice: *Stakeholders access, share, and apply successful PtD practices.*

    Practice focuses on identifying and sharing successful procedures, processes, equipment, tools, and results through on-line databases and other media. Practice also includes demonstrating the value of including workers' health and safety considerations in design decisions and exploring links with the movements toward "green" and sustainable design. The move in America, and elsewhere, towards more environmentally friendly practices is resulting in changes to traditional jobs and the creation of new kinds of occupations. As we make technological advances in industry, we need to remain vigilant in protecting workers against emerging hazards. As traditional jobs evolve to meet new challenges, workers may be faced with known risks that had not previously affected their occupation. These changes may also present us with the opportunity to eliminate hazards through planning, organization, and engineering.

4. **Policy:** *Business leaders, labor, academics, government entities, and standard-setting organizations endorse a culture that includes PtD principles in all designs affecting workers.*

> Policy focuses on creating demand for safe designs for workers and incorporating these safety and health considerations into guidance, regulations, recommendations, operating procedures, and standards.

5. **Small Business:** *Small businesses have access to PtD resources that are designed for or adapted to the small business environment.*

> These goals explore methods for tailoring and diffusing successful PtD practices and programs to small businesses.

Businesses and organizations are encouraged to learn more about methods to eliminate workplace hazards and minimize risks from PtD organizational partners (listed on page 16 of this Plan). Industrial sector-specific information can also be obtained from the NORA (National Occupational Research Agenda) Sector Programs:

- **Agriculture, Forestry, and Fishing:** http://www.cdc.gov/niosh/programs/agff/
- **Construction:** http://www.cdc.gov/niosh/programs/const/
- **Healthcare and Social Assistance:** http://www.cdc.gov/niosh/programs/hcsa/
- **Manufacturing:** http://www.cdc.gov/niosh/programs/manuf/
- **Mining:** http://www.cdc.gov/niosh/programs/mining/
- **Oil and Gas Extraction subsector:** http://www.cdc.gov/niosh/programs/oilgas/
- **Services:** http://www.cdc.gov/niosh/programs/pps/
- **Transportation, Warehousing, Utilities:** http://www.cdc.gov/niosh/programs/twu/
- **Wholesale and Retail Trade:** http://www.cdc.gov/niosh/programs/wrt/

Information on methods to eliminate specific hazards and/or minimize their associated risks can be obtained from the relevant NIOSH Cross-Sector programs. A link to a list of the NIOSH Cross-Sector programs is included here: http://www.cdc.gov/niosh/programs/.

In addition, Prevention through Design resources can also be found at the following websites:

- **Design for Construction Safety:** http://www.designforconstructionsafety.org
- **The Center for Construction Research and Training:** http://www.cpwr.com/

The PtD National Initiative is extensive and complex. A systematic, step-by-step approach to the completion of its activities (as outlined in this document) ensures that businesses and organizations have access to resources they need to implement successful PtD programs.

- The first step in the initiative is to create awareness of PtD in all industrial sectors, in organizations and agencies, and at schools and universities by sharing PtD benefits, success stories, and financial returns.

- Second, commitment must be obtained from businesses, organizations, and workers to incorporate PtD into work processes and health and safety management systems. Assistance will be needed to help stakeholders transition from traditional retrofitting activities and risk assessment and control methods to systems that include prevention considerations in all designs that impact workers. As part of this effort, policies are needed to facilitate the sustainable implementation of these changes in the workplace.

- Lastly, the impact of PtD on eliminating hazards and minimizing risks will be monitored to ensure that appropriate program and process adjustments occur as needed.

Because of the national scope of the PtD initiative, no single organization, occupational discipline, government agency, or individual holds the key to achieving its goals. All stakeholders must work collectively to meet the important objectives outlined in this report. We invite you to review the document in its entirety, offer comments by email to ptd@cdc.gov, and identify areas where you can make an impact, either within your own business or organization, or on behalf of the national initiative. To learn more about PtD or to express interest in participating, contact NIOSH by email at ptd@cdc.gov or by mail at 4676 Columbia Parkway, Mail Stop C–32, Cincinnati, OH 45226.

PREVENTION THROUGH DESIGN—RATIONALE

One of the best ways to prevent occupational injuries, illnesses, and fatalities is to eliminate hazards and minimize risks early in the design or redesign process and incorporate methods of safe design into all phases of hazard and risk mitigation. Although there is a long history of designing for safety for the general public in the United States, less attention has gone to factoring the safety, health, and well-being of workers into the design, redesign, and retrofit of new and existing workplaces, tools and equipment, and work processes. The National Institute for Occupational Safety and Health (NIOSH), building on a strong historical base, currently leads a nationwide initiative called Prevention through Design (PtD) [Schulte et al. 2008]. PtD addresses occupational safety and health needs by eliminating hazards and minimizing risks to workers *throughout the life cycle* of work premises, tools, equipment, machinery, substances, and work processes, including their construction, manufacture, use, maintenance, and ultimate disposal or reuse. A growing number of business leaders recognize PtD as a cost-effective means to enhance occupational safety and health. Many U.S. companies openly support PtD concepts and have developed management practices to implement them. PtD utilizes the traditional hierarchy of controls by focusing on hazard elimination and substitution, followed by risk minimization through the application of engineering controls and warning systems applied during design, redesign, and retrofit activities. However, PtD also supports the application of administrative controls and personal protective equipment when they supplement or complement an overall risk minimization strategy and include the appropriate program development, implementation, employee training, and surveillance.

Despite progress in PtD among individual businesses, the need for a PtD National Initiative remains compelling. In 2008, among U.S. workers, 5,071 died from occupational injuries, 3.7 million suffered serious injuries, and 187,400 became ill from work-related exposures [BLS 2008]. The annual direct and indirect costs of work-related injuries, illnesses, and fatalities have been estimated to range from $128 billion to $155 billion [Schulte 2005]. A recent study in Australia implicates design as a significant contributor to 37% of work-related fatalities [Driscoll et al. 2008]. Evidence suggests that the successful implementation of prevention through design concepts can greatly improve worker health and safety [NIOSH 2006; Lin 2008; Schulte et al. 2008].

The persistence in the United States of a large occupational morbidity, mortality, and injury burden demonstrates the need for a more concerted effort to reduce workplace risks than has been attempted in the past. The strategic plan outlined in this document establishes goals for the successful implementation of the PtD National Initiative. This comprehensive approach, which includes worker health and safety in all aspects of design, redesign, and retrofit, will provide a vital framework for saving lives and preventing work-related injuries and illnesses.

PROGRAM MISSION

The mission of the PtD National Initiative is **to prevent or reduce occupational injuries, illnesses, and fatalities through the inclusion of prevention considerations in all designs that impact workers.** The mission can be achieved by:

- Eliminating hazards and controlling risks to workers to an acceptable level *at the source* or as early as possible in the life cycle of equipment, products, or workplaces.

- Including *design, redesign, and retrofit* of new and existing work premises, structures, tools, facilities, equipment, machinery, products, substances, work processes, and organization of work.

- Improving worker safety and health through the inclusion of prevention methods in all designs that impact workers and others on the premises.

Strategic Agenda Development

The PtD National Initiative was launched during the first PtD Workshop, held in Washington, DC, on July 9–11, 2007. The purpose of the workshop was to engage stakeholders in open discussions about the need for eliminating occupational hazards and controlling risks to workers "at the source," or as early as possible in the life cycle of facilities, equipment, production, or processes. The PtD Workshop utilized the sector-based framework established by NIOSH for the second decade of the The National Occupational Research Agenda (NORA) (see Figure 1).

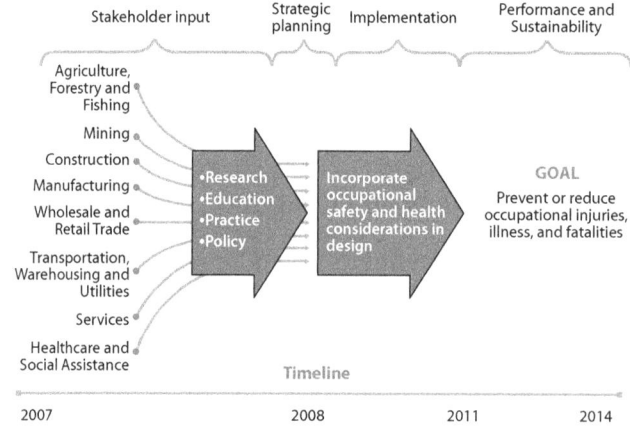

Figure 1. PtD National Initiative Timeline

The National Occupational Research Agenda (NORA)

The National Occupational Research Agenda (NORA) is a partnership program to stimulate innovative research and improved workplace practices. Unveiled in 1996, NORA has become a research framework for NIOSH and the nation. Diverse parties collaborate to identify the most critical issues in workplace safety and health. Partners then work together, developing goals and objectives to address these needs. Participation in NORA is broad, including stakeholders from universities, large and small businesses, professional societies, government agencies, and worker organizations.

NORA entered its second decade with a new sector-based structure to better move research to practice within workplaces. The national agenda for this second decade is being developed and implemented through NORA Sector Councils. Stakeholders within these Sector Councils are developing occupational safety and health strategic agendas for the nation within each of the following eight sectors: Agriculture, Forestry, and Fishing; Construction; Health Care and Social Assistance; Manufacturing; Mining; Services; Transportation, Warehousing, and Utilities; and Wholesale and Retail Trade.

Approximately 250 participants from diverse industry sectors and occupational disciplines attended the three-day event. Keynote and plenary speakers spotlighted the success of PtD in several industries in the United States and internationally. Participants engaged in breakout sessions to identify industry-centered opportunities and barriers, to develop sector-specific recommendations, and to map out overarching issues for the four functional areas in the PtD framework (*Research, Education, Practice, and Policy*). Reporters from the eight industrial sectors and four functional areas summarized the discussions of their breakout sessions and published their findings in the April 2008 edition of the *Journal of Safety Research* [Bealko et al. 2008; Behm 2008; Chapman and Husberg 2008; Fisher 2008; Gambatese 2008; Heidel 2008; Howe 2008; Johnson 2008; Lin 2008; Madar 2008; Mann 2008; Mroszczyk 2008]; [http://www.cdc.gov/niosh/topics/PtD/PTDJournalArticles.html].

The Workshop was hosted by NIOSH and planned in collaboration with the following organizations: American Industrial Hygiene Association, American Society of Safety Engineers, The Center for Construction Research and Training, Kaiser Permanente, Liberty Mutual, National Safety Council, Occupational Health and Safety

Administration, ORC Worldwide (now Mercer), and Regenstrief Center for Health Care Engineering. These organizations endorse and support Prevention through Design and are currently engaged in activities to meet the goals of the National Initiative. These organizations, along with the Association of Equipment Manufacturers, serve as NIOSH PtD organizational partners. Other partners are welcome to join.

A second PtD workshop, titled *Making Green Jobs Safe*, attended by 170 green economy stakeholders, was held in Washington, DC, on December 14–16, 2009, to frame the issues for ensuring that green jobs, technologies, and products are safe for workers. Stakeholders at this workshop developed a prioritized list of activities, that, when implemented, will result in the inclusion of occupational health and safety goals into the social aspects of sustainable design.

Subsequent to the workshop in 2007, a PtD Council was convened consisting of a broad range of stakeholders, including organizational partners, representatives from industry, labor, academia, and government, and PtD experts. The PtD Council members agreed to help draft the Plan for the National Initiative on PtD, including specific goals, activities, outcomes, and performance measures for successfully integrating PtD into all business practices that affect workers. PtD Council members will also participate in ongoing meetings and lead workgroups to implement the PtD agenda. Corresponding members of the PtD Council consist of interested stakeholders who are willing to participate in any of the specific work groups.

The ideas and discussions that came out of the PtD workshops (along with input from the eight NORA Sector Councils) serve as the basis for the Plan for the National Initiative on PtD. The plan provides a road map for the nation for integrating prevention considerations into all designs that affect the health and safety of workers. The eight NORA Sector Councils, in conjunction with the PtD Council, will identify sector-specific goals that can be accomplished by implementing enhanced prevention-based designs. The plan will be aligned with the strategic agendas being developed for the eight NORA sectors. This alignment will help ensure that the PtD goals become integral to the sector agendas. Many issues raised at the sector level will apply to more than one industry. The plan will capture these issues within the cross-sector functional areas of *Research, Education, Practice, and Policy*.

The success of the Plan for the National Initiative on PtD depends on: (1) members' commitment to participating in a particular goal and/or their desire to serve as a partner in any dissemination efforts; (2) the availability of research funding and interested researchers; and (3) the availability of individuals with PtD expertise to help implement the program. NIOSH will act as coordinator for the plan. Ultimately, its success depends on active participation by partners and other stakeholders who are charged with working to accomplish the goals sets forth in the plan.

For more information about PtD partner organizations, please visit their Web sites:

— American Industrial Hygiene Association: http://www.aiha.org/
— American Society of Safety Engineers: http://www.asse.org/
— Association of Equipment Manufacturers: http://www.aem.org/
— The Center for Construction Research and Training: http://www.cpwr.com/
— Kaiser Permanente: https://www.kaiserpermanente.org/
— Liberty Mutual: http://www.libertymutualgroup.com/omapps/Content Server?pagename=LMGroup/Views/LMG&ft=3&fid=1138364006146&ln=en
— National Safety Council: http://www.nsc.org/
— Occupational Safety and Health Administration: http://www.osha.gov/
— ORC Worldwide™: http://www.orc-dc.com/
— Regenstrief Center for Healthcare Engineering: http://www.purdue.edu/dp/rche/

Because of the national scope of the PtD initiative, no single organization, occupational discipline, government agency, or individual holds the key to achieving its goals. All stakeholders must work collectively to meet the important objectives outlined in this report. We invite you to review the document in its entirety, offer comments by email to ptd@cdc.gov, and identify areas where you can make an impact, either within your own business or organization, or on behalf of the national initiative. To learn more about PtD or to express interest in participating, contact NIOSH by email at ptd@cdc. gov or by mail at 4676 Columbia Parkway, Mail Stop C–32, Cincinnati, OH 45226.

Case Study 1—Prevention through Partnerships: Industry-wide Equipment Design Changes Protect Workers' Health

More than half a million workers in the United States face exposure to fumes from asphalt, a petroleum product used extensively in road paving, roofing, siding, and concrete work. Health effects of concern from exposure to asphalt fumes include headache, skin rash, sensitization, fatigue, reduced appetite, throat and eye irritation, cough, and lung and skin cancer.

In the mid-1990s, The National Asphalt Pavement Association (NAPA) collaborated with industry partners, labor, and NIOSH to address concerns about the hazards of asphalt fumes for paving-site workers. This partnership resulted in U.S. industry-wide efforts to design, lab-test, field-test, and validate engineering control modifications to highway-class paving machines to remove fumes from the vicinity of workers. The industry partners signed a voluntary agreement with OSHA to include these ventilation controls into every U.S.-manufactured highway-class paver.

[Cervarich 2008]

Before and after photos of asphalt fume emissions from highway-class pavers

Areas for Goals Development

The goals within the PtD agenda are organized around four overarching areas: *Research, Education, Practice, and Policy. Small Business* was added as an additional focus for goal development to address the unique challenges of applying PtD methods to small business operations, processes and environments. Each of these overarching areas, as well as the small business focus area, is supported by a strategic goal. Details about specific activities for accomplishing each goal, as well as performance measures and timeframes, can be obtained by clicking on the specified links.

1. **Research:** *Research will establish the value of adopted PtD interventions, address existing design-related challenges, and suggest areas for future research.*

   Research focuses on design factors that effectively reduce occupational morbidity and mortality; metrics that assess the impact of these design factors; methods that diffuse effective designs; and, economic and business issues.

   Research topics address the following questions:

   - Do we have evidence to support designs that will effectively eliminate hazards and reduce risks to workers?
   - Do we understand the role design plays in each sector's serious injury, illness, fatality, and exposure experience?
   - What are the motivators for, and barriers to, effective implementation of PtD?
   - Can we define effective business cases for implementing PtD?
   - How do we influence a culture change toward designing for worker safety and health?

   Questions such as these prompt the need for specific research by qualified investigators at universities, research institutes, organizations, and corporations. The answers will further the efforts being conducted under the education, practice, and policy goals.

2. **Education:** *Designers, engineers, machinery and equipment manufacturers, health and safety (H&S) professionals, business leaders, and workers understand PtD methods and apply their knowledge and skills to the design and redesign of new and existing facilities, processes, equipment, tools, and organization of work.*

   Education focuses on motivating and equipping professional communities to continually increase their capacity to identify health hazards and assess and minimize risks for worker safety and health. Acquisition of PtD expertise will be acquired through enhanced design and engineering curricula as well as through improved professional accreditation programs that value PtD knowledge and skills and include them in their competency assessments. Making business leaders aware of the potential for increasing company profitability by incorporating prevention through design methods into their systems and processes is also an important component of the education goal.

PtD requires the development and implementation of a broad educational framework adapted to the full range of occupational disciplines and educational settings involved in supporting the PtD initiative. The educational objectives and content will vary significantly according to the individual discipline or education setting. For example, the educational needs of the mechanical engineer will vary significantly from those of the architect, industrial designer, purchasing agent, or finance professional. However, there are common PtD themes that will be woven into this broad educational framework. Fundamental to the concept of risk management is the ability to accurately assess risk and recommend acceptable risk levels through the application of the hierarchy of controls. Educating professionals in all design-related occupational disciplines on the determination of acceptable risk and subsequent communication to management is vital to the success of PtD. Furthermore, the educational framework must address the needs of students at the beginning of their careers, as well as those of experienced professionals. For students, PtD educational material could be integrated into existing courses, textbooks, and certifications. For experienced professionals, PtD concepts could be delivered via professional development courses, continuing education seminars, and journal publications. PtD knowledge, skills, and information must also reach small business owners who may not have access to professional development courses. This may be accomplished by utilizing various business and trade associations as well as through product and service suppliers. Finally, but no less important, PtD education must reach the worker. Novel educational formats, such as Web-based video clips, could broaden our reach to workers in all industrial sectors and in large and small workplaces.

3. **Practice:** *Stakeholders access, share, and apply successful PtD practices.*

Practice focuses on identifying and sharing successful procedures, processes, equipment, and tools through on-line databases and other media. Practice also includes demonstrating the value of including workers' health and safety in design decisions and exploring links with the movement towards sustainability.

The effort that will be required to implement the successful practice of PtD should not be underestimated. Examples of some actions that will be needed to promote the practice of PtD include:

- Developing business cases that are compelling to business leaders (see Case Study 2);
- Accurately assessing risks inherent to various designs, processes, or procedures; and
- Designing successful tools and equipment (e.g., fall protection systems or machine guarding).

Essential to the practice of PtD will be the development of a web-based system that utilizes standardized evaluation criteria to share successful PtD processes, procedures, tools, and equipment. Providing methods for identifying tools and equipment that include safety and health design elements will impart valuable guidance not only to businesses, but also to consumers.

Workers also play a key role in the success of the current initiative. Their input on the creation of safe work areas, tools, and tasks is critical to shaping successful design systems. Identifying successful programs that include worker input in the design process should play a central role in the PtD agenda.

Given current economic realities, it is essential to demonstrate the value of PtD in lean manufacturing to ensure that the implementation of the lean approach does not result in the elimination of programs essential to worker health and safety. By applying PtD methods, companies can not only reduce the potential for worker injuries and illnesses (and mitigate the costs associated with workplace injuries and illnesses) but also increase their profits through improved work processes and work organization. Developing and disseminating case studies that showcase companies that improved profits through Prevention through Design methods will provide valuable guidance to other businesses.

4. Policy: *Business leaders, labor, academics, government entities, and standard-developing and -setting organizations endorse a culture that includes PtD principles in all designs affecting workers.*

Policy focuses on creating demand for safe designs for workers and incorporating these safety and health considerations into guidance, regulations, recommendations, operating procedures, and standards.

Most urgent to the implementation of PtD is the development of a broad, overarching policy that will serve as a roadmap for establishing

Case Study 2—Mechanical devices reduce risks to workers and improve patient safety

The ergonomic hazards associated with lifting present significant risk to health care workers in the United States. In 2003 alone, caregivers suffered 211,000 occupational injuries. As the population ages and the demand for skilled care services continues to rise, the incidence of musculoskeletal injuries to the back, shoulder, and upper extremities of caregivers will also increase. Lifting without the assistance of mechanical devices can also compromise the safety and comfort of patients.

A NIOSH study evaluated the effectiveness of a safe resident lifting and movement intervention in six nursing homes. After investment of $158,556 for patient lifting and handling equipment and worker training, lost workday injuries fell by 66%, restricted workdays dropped 38%, and workers' compensation expenses were reduced 61%. The initial investment for the lifting and handling equipment and worker training was recovered in less than three years, based on post-intervention savings of $55,000 annually in workers' compensation costs.

[NIOSH 2006]

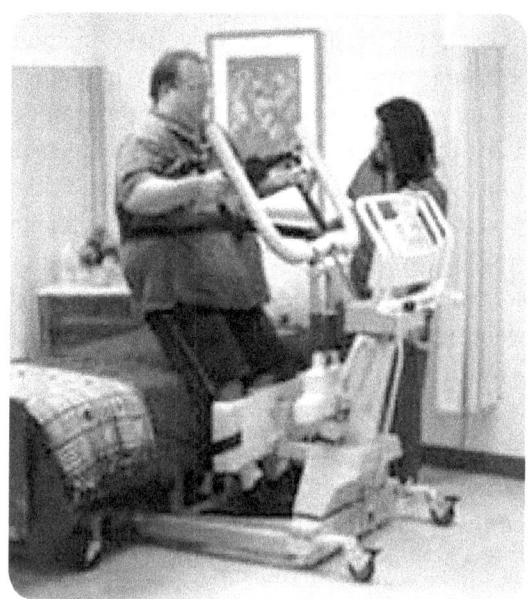

Preventing Injury through Effective Design: Mechanical lifting devices reduce risk of back injuries to health care workers and improve patient safety and comfort.

PtD processes and programs for enterprises of all sizes, across all industrial sectors. The creation of outcome-based guidance for the implementation of industry- or activity-specific standards is also needed. As a fundamental element to the development of PtD policy, relevant recommendations from various authoritative and advisory organizations should reflect PtD principles. The ultimate goal is to include PtD principles in all design standards that affect workers.

5. **Small Business:** *Small businesses have access to PtD resources that are designed for or adapted to the small business environment.*

The Small Businesses goals focus on methods for tailoring and diffusing successful PtD practices and programs to small businesses. Since small businesses must operate effectively on limited resources, they often do not have the in-house capacity to address prevention through design issues. Thus, the national initiative calls for adapting successful PtD practices and programs to the small business environment. Case studies highlighting successful practices, programs, and interventions at the small business level could help promote PtD methods to these establishments.

Figure 1 provides a timeline for the PtD National Initiative. The first phase of the initiative, in July 9–11, 2007, was to gain stakeholder input on the goals of the initiative. During the second phase, stakeholder input was analyzed and the strategic and intermediate goals and activities to achieve these goals were developed. In 2009, the implementation phase began. This phase involves creating awareness about PtD in all industrial sectors; planning for the significant changes organizations will make in the way they manage health and safety; and implementing the main elements of the PtD National Initiative. Implementation will include steps to design and conduct various tasks. It will also involve the identification of indicators and measures for assessing success of those tasks and their larger objectives. In the fourth year of the PtD National Initiative (2011), a conference will be held to take stock of achievements accomplished to date and to identify the aspects of the initiative and its implementation that require further effort or enhancements. In the three years subsequent to the conference (2012 to 2014), stakeholders will continue to implement the national initiative with the inclusion of these additional efforts identified during the 2011 conference.

A summary of the activities that individual companies and organizations must complete to achieve each step in adopting a successful PtD program in their workplace is illustrated in Figure 2 on page 16. These activities include creating awareness of PtD; gaining commitment from PtD stakeholders; assessing PtD needs and developing plans for PtD adoption; implementing PtD plans and processes; and, monitoring performance and making changes as needed.

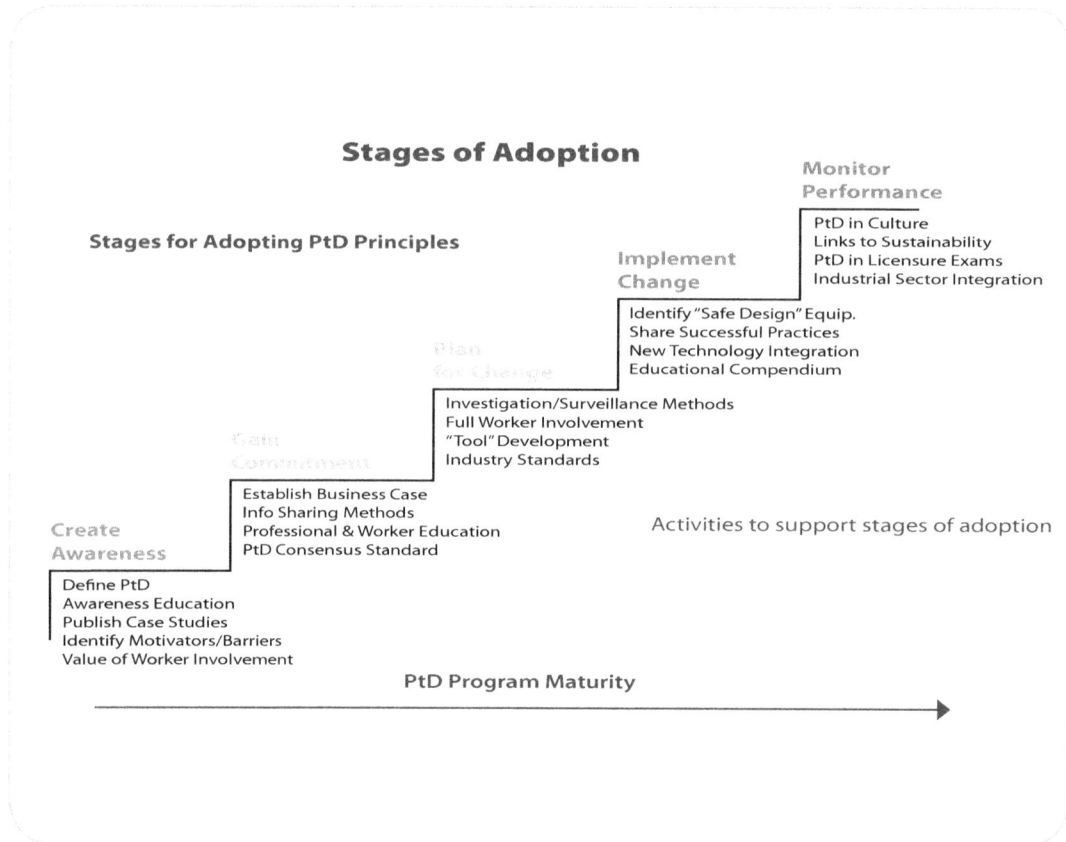

Stages of Adoption

Stages for Adopting PtD Principles

Monitor Performance
- PtD in Culture
- Links to Sustainability
- PtD in Licensure Exams
- Industrial Sector Integration

Implement Change
- Identify "Safe Design" Equip.
- Share Successful Practices
- New Technology Integration
- Educational Compendium

Plan for Change
- Investigation/Surveillance Methods
- Full Worker Involvement
- "Tool" Development
- Industry Standards

Gain Commitment
- Establish Business Case
- Info Sharing Methods
- Professional & Worker Education
- PtD Consensus Standard

Create Awareness
- Define PtD
- Awareness Education
- Publish Case Studies
- Identify Motivators/Barriers
- Value of Worker Involvement

Activities to support stages of adoption

PtD Program Maturity

Figure 2. PtD Stages of Adoption

Case Study 3— Designing-Out Existing Noise Hazard: Protecting Miners' Hearing While Experiencing a Positive Return on Investment

Overexposure to noise remains a widespread and serious health hazard to workers in many U.S. industries. Hearing loss is permanent and can lead to poor verbal communication and the inability to recognize warning signals. Workers who experience hearing loss can also suffer from increased stress and fatigue.

In underground coal mining, the operation of continuous mining machines—which cut and gather coal—poses a significant risk to miners' health. These machines contain an onboard conveyor consisting of a chain with flight bars that drag the coal along the base of the conveyor system. The chain creates excessive noise as it makes contact with the metal base and the coal. Mine operators working in proximity of the machine are at risk of permanent and irreversible hearing loss.

An innovative PtD solution calls for coating the chain conveyor and flight bars with a heavy-duty, highly durable urethane. The redesigned chain conveyor and flight bars decrease noise exposures of continuous mining machine operators by 3 dB(A), a significant reduction in noise exposure. This PtD redesign also increases the chain life by a factor of three, more than off-setting the 20% cost increase of coating the chain.

[Kovalchik et al. 2008]

Design improvements to the onboard conveyor of a continuous coal mining machine reduce noise exposures. Coating the chain conveyor and flight bars protects mine operators' hearing, and extends the life of the chain.

Case Study 4—Safety on the Seas: E-stop on Fishing Vessels Reduces Risk of Serious Injury or Death

Commercial fishing is the most dangerous occupation in the United States. Its occupational fatality rate—142 per 100,000 workers—is 36 times higher than the average fatality rate for all U.S. workers. Fatal deck injuries, caused by unsafe machinery and equipment, remain a significant problem in the industry.

In Alaska, deck injuries result in approximately 12% of all commercial fishing fatalities and 67% of severe injuries necessitating hospitalized. Entanglements—especially those associated with powerful, capstan-type winches used to reel in large fishing nets—pose a significant threat to deck workers in the commercial fishing industry.

The capstan winch is usually mounted in the center of the deck. The winch's drum rotates while the crew works on deck. Fishermen who lose their balance or are inattentive can become caught in the fishing line as it winds around the drum. The winch provides no entanglement protection and the controls are usually out of reach of the entrapped person.

A unique Prevention through Design solution—an emergency-stop ("e-stop") system situated on the top of the winch—addresses the serious machinery hazard posed by a capstan deck winch. When pushed, the emergency switch arrests the drum's rotation in less than 180 degrees rotation, sufficient to limit serious entanglement injury. Vessel owners who have the e-stop installed on their winches recommend it enthusiastically to other fishermen.

[Lincoln JM et al. 2008]

(A)   (B)

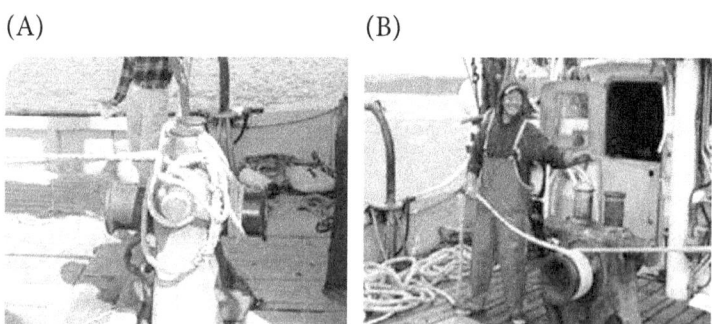

(A) A captan-type winch with fishing lines wound around. (B) A fishing captain demonstrating the use of an emergency-stop (e-stop) mounted on the winch.

Goal Development

The PtD Plan for the National Initiative includes five strategic goals, one for each of the four functional areas (*Research, Education, Practice, and Policy*) and one to address the unique business environment of *Small Business*. Each strategic goal is supported by one or more intermediate goals that organizations or individuals must undertake in support of the strategic goals. To help gauge progress towards meeting the desired outcome within the specified timeframe, a performance measure is specified for each intermediate goal. Finally, activities are defined, including outputs and transfers to stakeholders. These activities include the creation of tools, controls, guidelines, training materials, recommendations, new knowledge, surveillance systems, documents, policies, and conferences. Milestones that set the timeframes for these activities are also provided when known.

The strategic and intermediate goals for the PtD National Plan are provided on the following pages. Activities are specified to serve as guideposts and to identify gaps and opportunities in the PtD National Plan. The PtD National Initiative will coordinate PtD activities and leverage resources between and among the eight NORA sectors.

The large number of goals and activities attests to the size of the task to achieve a culture change that includes worker health and safety in aspects of design, redesign, and retrofit. While all of the goals and activities are important, the following are considered critical to the initiative:

1. Conduct research to characterize hazards, assess risks, and develop and implement control measures for new technologies, processes, and products prior to adoption;

2. Support the inclusion of PtD principles in consensus standards and guidance;

3. Influence the inclusion of PtD into the engineering process, core curriculum and accreditation, textbooks, and professional certification;

4. Achieve sustainability through the application of PtD to ensure that the resulting health, energy and environmental benefits can be at their highest levels for workers, the public, and the environment;

5. Train workers in their PtD role and ensure workers' input in eliminating hazards and minimizing risks associated with their job activities.

PtD National Initiative Goals

1. Research Goals

Strategic Goal: Research will establish the value of adopted PtD interventions, address existing design-related challenges, and suggest areas for future research.

1.1. Intermediate Goal 1: Support research to evaluate the effectiveness of design interventions.

Performance Measure: By 2011, evidence supporting the effectiveness of selected design interventions in preventing injuries, illness, fatalities, and exposures will be gathered and disseminated.

1.1.1 Activity/Output Goal: Survey ORC Worldwide™ companies to determine the level and results of adoption of PtD among member companies. This information will be analyzed to determine whether work-related injuries, illnesses, and fatalities in a group of U.S.-based companies have similar, design-related factors *(from the NIOSH/PtD NORA Research Project)*.

> "Two hundred ten occupational fatalities occurring between 2000 and 2002 in Australia were analyzed for design-related factors. Design factors played a definite or probable role in 37% of the cases. In another 14%, design-related factors may have played a contributing role."
>
> [Driscoll et al. 2008]

1.1.1.1. *Milestone: By 2010, disseminate the research findings to the U.S. manufacturing industry, via the NORA Manufacturing Sector, for application in practice.*

1.1.2 Activity/Output Goal: Compare and contrast ergonomic design interventions to determine the factors that facilitate successful design outcomes.

1.1.3 Activity/Output Goal: Review published research/literature for effective design solutions.

1.1.4 Activity/Output Goal: Investigate the effectiveness (of reducing injuries, illnesses, and exposures) of previously implemented interventions and publish the results.

1.1.5 Activity/Output Goal: Investigate the conversion of investigation and surveillance data to actual design solutions.

1.1.6 Activity/Output Goal: Investigate PtD effectiveness of ANSI Z10 2005 and determine opportunities for integrating PtD principles, as appropriate.

1.1.6.1 Milestone: By 2011, disseminate the research findings on PtD effectiveness of ANSI Z10 2005 and opportunities for integrating PtD principles.

1.1.7 Activity/Output Goal: Assess the effects of 12 years of PtD regulations on construction companies in the UK *(from NIOSH/PtD NORA Research Project)*.

1.1.7.1 Milestone: By 2011, disseminate the research findings to U.S. construction industry, via the NORA Construction Sector, for application in practice.

1.1.8 Activity/Output Goal: Identify and communicate design factors that are successful in reducing occupational motor vehicle fatalities, in automobiles as well as in commercial vehicles.

1.1.9 Activity/Output Goal: Investigate the contributions of vehicle selection, standard and optional equipment, maintenance, and inspection on occupational motor vehicle accident rates.

1.1.10 Activity/Output Goal: Investigate the effects of distracted, fatigued, and impaired drivers on occupational motor vehicle accidents. Determine and communicate effective design interventions.

1.1.11 Activity/Output Goal: Present results of motor vehicle design improvements to the top 5 (five) opportunity companies in the Wholesale and Retail Trade (WRT), Services, and Transportation, Warehousing, and Utilities (TWU) NORA sectors.

1.1.12 Activity/Output Goal: Determine the level of adoption of PtD concepts among a subset of Fortune 500 companies in the United States through a survey conducted by ORC Worldwide *(from the NIOSH/PtD NORA Research Project)*.

1.1.12.1 Milestone: By 2011, complete guidelines for implementation of PtD concepts within the corporate setting.

1.1.13 Activity/Output Goal: Investigate the design implications for vulnerable worker populations, including workers with disabilities and aging, immigrant, and overweight workers.

1.1.14 Activity/Output Goal: Establish mechanisms to ensure pro-active research on OSH hazards and risks for new products and technologies.

1.2. Intermediate Goal 2: Improve existing surveillance techniques to better identify design-related factors

Performance Measure: By 2011, sector-specific incident investigation and reporting models that include design factors will be proposed and dissemination plans developed. Mechanisms to receive reports of design-related factors that can influence worker health and safety will be proposed, including dissemination plans.

1.2.1   Activity/Output Goal: Investigate mechanisms to determine market penetration of safer designs. Include legislative, litigation, insurance, and market-driven mechanisms.

1.2.2   Activity/Output Goal: Investigate effective methods for reporting design-related incidents (e.g., "Report Unsafe Products" program sponsored by the Consumer Product Safety Commission) to determine their use as models for reporting of design-related factors involving occupational injuries/illness/fatalities or their potential. Review successful existing surveillance instruments for linking injuries, illnesses, and exposures to design-related factors.

1.2.3   Activity/Output Goal: Investigate surveillance mechanisms, including benchmarking forums and trade association surveys to effectively capture design-related worker health and safety incidents.

1.2.4   Activity/Output Goal: Derive research questions that would drive Bureau of Labor Statistics (BLS) data collection of design-related injuries, illnesses, and fatalities. Integrate PtD surveillance needs into BLS data collection (see Intermediate Goal 3.2.1).

1.2.5   Activity/Output Goal: Review existing incident investigation models (e.g., published construction models) that identify causal factors. Recommend improvements that include specific criteria for the identification and reporting of design-related factors. Adapt models to other industrial sectors and ensure that they are valid and reliable.

1.2.6   Activity/Output Goal: Create a surveillance program for green transport that includes the development of early indicators.

1.2.7   Activity/Output Goal: Use migrant health clinics/extension service for sentinel event alerts.

1.2.8   Activity/Output Goal: Include occupational injury and illness tracking component in health records.

1.3.   *Intermediate Goal 3: Investigate the relationship between the "hierarchy of controls" and the business value of PtD.*

Performance Measure: By 2011, compelling business cases for including PtD methods into workplace designs are investigated and communicated. Return-on-investment tools and business case models are evaluated and incorporated into the PtD process.

1.3.1   Activity/Output Goal: Evaluate American Industrial Hygiene Association (AIHA) case studies as part of the "Value of the Profession" project for applicability to PtD. Include links to major business goals and investments. Include findings in a white paper.

1.3.2   Activity/Output Goal: Evaluate existing financial analysis tools and calculators; investigate need for modifications or updates. Broaden measures of costs to include more indirect costs/benefits. Determine the appropriate methods to incorporate financial models into the overall PtD process.

   1.3.2.1   *Milestone: By 2011, recommend appropriate methods for calculating financial indicators.*

1.3.3   Activity/Output Goal: Identify and communicate the compelling story for each sector, linking value to business.

   1.3.3.1   *Milestone: By 2011, develop and disseminate a library of sector-specific business case studies for all eight industrial sectors.*

1.3.4   Activity/Output Goal: Investigate the impact of and alignment with "Lean Manufacturing" in achieving productivity and quality improvements. Identify case studies linking productivity, quality, process efficiency, and safety.

1.3.5   Activity/Output Goal: Develop business cases for both successfully adopted and under-utilized design solutions (two adopted and four to six under-utilized). Promote widespread implementation of underused design solutions that are supported by strong business cases by applying diffusion strategies (from the NIOSH/PtD NORA Research Project).

   1.3.5.1   *Milestone: Business case development by 2011. Engineering control strategies disseminated to target industry by 2012.*

1.4.   *Intermediate Goal 4: Investigate the motivators, enablers, and barriers to PtD implementation.*

Performance Measure: By 2009, knowledge of PtD motivators, enablers, and barriers will be incorporated into the PtD activities. Plans will be developed to benefit from the motivators and enablers, resolve concerns, and break through the barriers.

> *"Proposed health and safety investments in preventive design may not be implemented without either sound financial return coupled with solutions that are proven to reduce occupational injuries, illnesses or fatalities or a full realization of the cost implications. Health and Safety professionals require financial return on investment models and tools so that they can include financial considerations in their capital requests. In addition, "designing out" hazards and risks may be associated with improvements in productivity."*
>
> [Gambatese JA 2008]

1.4.1　Activity/Output Goal: Explore the American Institute of Architects (AIA) position on including safe construction methods into designs.

1.4.2　Activity/Output Goal: Investigate product liability and completed operations aspects of including elements to reduce risk of operator injury or illness into tool and equipment design.

1.4.3　Activity/Output Goal: Facilitate communications between architecture and engineering firms and the appropriate insurance carriers to develop a risk management plan for facility design and materials to make buildings safer to construct, maintain, and operate.

> *"Liability for effects of designs is perceived by some to be a barrier to wider use of PtD. Issues of liability, with regard to PtD, relate to both the designer and final owner, and are handled in various ways in various countries."*
>
> *"...further studies are needed to determine whether designers feel a sense of liability when designing tools or projects, and whether that perception is real or only perceived. The barrier to the use of liability as a motivator is the notion that it could sometimes be detrimental to innovation."*
>
> [Lin 2008]

1.4.4　Activity/Output Goal: Gain feedback on motivators, enablers, and barriers to the successful implementation of PtD from PtD organizational members, such as American Industrial Hygiene Association (AIHA), the American Society of Safety Engineers (ASSE), National Safety Council (NSC), and ORC WorldWide™.

1.5.　Intermediate Goal 5: Investigate the relationship among organizational safety culture, design team-member roles, and effective implementation of PtD methods.

Performance Measure: By 2011, investigate and define the role changes that are required to effectively implement PtD.

1.5.1　Activity/Output Goal: Investigate persons, and their respective roles, involved in task-based and user-based design. Publish a white paper and/or present at relevant conferences.

1.5.2　Activity/Output Goal: Investigate successful PtD programs at three organizations to determine the mechanics for implementation of PtD as well as roles and responsibilities of involved persons.

2. Education Goals

Strategic Goal: Designers, engineers, machinery and equipment manufacturers, health and safety (H&S) professionals, business leaders, and workers understand PtD methods and apply their knowledge and skills to the design and redesign of new and existing facilities, processes, equipment, tools, and organization of work.

2.1   Intermediate Goal 1: Develop/disseminate educational programs, training modules, and case studies, including:

— *Engineering, architecture, business, health and safety professional, and allied health curricula;*

— *Vocational-technical programs;*

— *Continuing education courses, professional development courses, and seminars;*

— *Industry, trade, and public service organization courses;*

— *Training modules for workers and employers.*

"Educating engineers, architects and designers about prevention through design concepts will require a multi-faceted approach. Including PtD courses into the core curriculum of universities will not be possible; rather, weaving PtD concepts and case studies into existing engineering and design courses is a more practical solution to the need for PtD education at the university level."

—[Mann JA 2008]

2.1.1   Activity/Output Goal: Enlist the support of deans of engineering schools to include basic PtD and occupational safety and health principles in required engineering courses.

2.1.1.1   *Milestone: By 2012, 25% of deans contacted agree to include PtD methods and tools in required engineering courses.*

2.1.2   Activity/Output Goal: Integrate "safe and lean" concepts, such as those outlined in ANSI B11 TR7 on Designing for Safety and Lean Manufacturing, into the Lean Manufacturing engineering curricula.

2.1.3   Activity/Output Goal: Include PtD concepts and methods into curricula at NIOSH Education and Research Centers (ERCs) and Training Project Grants (TPGs) (*from NIOSH PtD Program*).

2.1.3.1   *Milestone: By 2012, 50% of ERCs and TPGs include PtD concepts and methods in curricula.*

2.1.4   Activity/Output Goal: Work with the National Institute of Standards and Technology (NIST) on integrating PtD principles into the Hollings Manufacturing Extension Partnership (MEP) program (*from NIOSH PtD Program*).

2.1.4.1   *Milestone: By 2012, Prevention through Design topics are available to Small Businesses through the NIST MEP program.*

2.1.5   Activity/Output Goal: Provide technical educational seminars and presentations to engineering, architecture, and environmental organizations (e.g., National Association of Environmental Managers).

   2.1.5.1 *Milestone: PtD awareness reach is expanded to three (3) new non-health and safety (H&S) organizations each year, from 2009–2014.*

2.1.6 Activity/Output Goal: Include PtD professional development courses, roundtables, and presentations at all major H&S conferences (from NIOSH PtD Program).

   2.1.6.1 *Milestone: PtD content is reinforced at all major H&S conferences from 2009–2014.*

2.1.7 Activity/Output Goal: Educate H&S professionals on demonstrating the business value of PtD projects. Develop and conduct a "Return on Investment" professional development course at AIHA, ASSE, and NSC beginning in 2010. Develop a brief publication on the topic. Develop computer-based training or podcast versions.

   2.1.7.1 *Milestone: By 2010, H&S professionals receive a framework and tools for demonstrating the business value of PtD projects as well as examples.*

2.1.8 Activity/Output Goal: Include cultural and linguistic considerations and accommodations in the design of worker H&S education programs.

   2.1.8.1 *Milestone: Beginning in 2011, H&S education programs include review by cultural and linguistic experts to ensure appropriate considerations and accommodations are included.*

2.1.9 Activity/Output Goal: Review National Safety Council (NSC) Institute for Safety through Design educational compendium to determine supplemental information that can be provided or to increase the dissemination of the existing programs.

   2.1.9.1 *Milestone: By 2011, complete review of the NSC Institute for Safety through Design educational compendium and issue report.*

2.1.10 Activity/Output Goal: Include PtD in 10-hour courses at the OSHA Training Institute.

   2.1.10.1 *Milestone: By 2012, PtD concepts will be incorporated into the OSHA 10-hour course at the OSHA Training Institute.*

2.1.11 Activity/Output Goal: Develop an inventory and summary of existing PtD course materials and training programs at universities and through H&S organizations.

   2.1.11.1 *Milestone: By 2011, publish inventory of existing PtD course materials and training programs delivered at Universities and through H&S organizations.*

2.1.12 Activity/Output Goal: Incorporate PtD into the criteria for accrediting engineering, engineering technology, and applied science programs by the Accreditation Board for Engineering and Technology (ABET).

> *2.1.12.1 Milestone: By 2011 PtD principles are incorporated into the criteria for ABET accreditation.*

2.1.13 Activity/Output Goal: Include PtD principles in the National Council of Examiners for Engineering and Surveying (NCEES) licensure exams.

> *2.1.13.1 Milestone: By 2011, PtD principles are included in the NCEES licensure exams.*

2.1.14 Activity/Output Goal: Include PtD concepts in the National Sciences Resource Center (NSRC) Science Awareness Program.

> *2.1.14.1 Milestone: By 2012, PtD concepts are included in the NSRC Science Awareness Program.*

2.1.15 Activity/Output Goal: Demonstrate how PtD concepts can be diffused to engineering school curricula by incorporating the relevant case studies into engineering textbooks that focus on design.

> *2.1.15.1 Milestone: PtD concepts and/or case studies will be included in seven (7) new editions of engineering textbooks by 2011.*

2.1.16 Activity/Output Goal: Work with engineering educators and PtD safety engineering experts to develop a catalog of PtD knowledge topic areas and related skills for the engineer, and identify available courses or other sources of content.

2.1.17 Activity/Output Goal: Identify risk assessment methods essential to the establishment of acceptable risk levels through the application of the hierarchy of controls and include these methods in PtD educational programs, training modules, and case studies.

2.1.18 Activity/Output Goal: Train new U.S. workforce to operate green technologies safely.

2.1.19 Activity/Output Goal: Educate the public on recycling and the collection process through public and community outreach programs.

2.1.20 Activity/Output Goal: Educate practitioners and emphasize importance of using upstream thinking (planning, preventing, controlling) as the primary approach to including OSH into sustainability.

2.1.21 Activity/Output Goal: Determine how to most quickly train workers for performing green jobs safely (e.g., integrating OSH into community college retraining programs).

2.1.22 Use Susan Harwood Grants for green worker training.

2.2 *Intermediate Goal 2: Educate decision-makers about the value of PtD in the design, redesign and retrofit of facilities, processes, equipment, tools, and organization of work, as well as the value of purchasing equipment and tools that incorporate safe design features.*

Performance Measures: Beginning in 2011, extend reach of the value of PtD beyond H&S and engineering professionals to include business management, finance, and purchasing professionals through classroom courses, journal publications, white papers, and conference presentations.

2.2.1 Activity/Output Goal: After the spring 2009 launch of the NIOSH and Xavier University MBA course on the "Business Value of Safety and Health" (developed in partnership with the National Safety Council to address how business can use occupational safety and health improvements and initiatives for long-term planning, operations management, and other decision-making), determine the potential for sharing key elements of this course through online courses, computer-based training, and webinars.

2.2.2 Activity/Output Goal: For each industrial sector, identify the decision-makers who can influence the adoption of PtD methods in their respective companies or organizations, the journals or publications that are relevant to their profession and industry, their trade associations, and the conferences they attend. Use this output for Activity/Output Goal 2.2.4

2.2.3 Activity/Output Goal: Develop a PtD message for each NORA industry sector, outlining the business value of PtD in the design and redesign of facilities, processes, equipment, tools, and organization of work as well as the value of purchasing equipment and tools that incorporate safe design features. Disseminate the messages via the NORA industrial sectors.

2.2.4 Activity/Output Goal: Develop a plan to increase our PtD reach to these decision-makers through their preferred journals and publications, trade associations and/or conferences. Work with the NORA sectors to implement the plan.

2.3 *Intermediate Goal 3: Influence business leaders and administrators to include safety design education and experience in engineering and architecture position descriptions.*

Performance Measure: By 2011, business leaders and administrators recognize the value of PtD knowledge and skills for their engineering and design staff members and include PtD credentials in position descriptions.

2.3.1 Activity/Output Goal: Identify the top 500 design/engineering firms in the United States. Develop a message to influence these firms to include PtD principles in their designs.

> 2.3.1.1 *Milestone: Twenty-five percent (25%) of contacted firms request additional information about PtD.*

2.3.2 Activity/Output Goal: Identify and influence engineering firms recognized for their "sustainable design" activities to include PtD skills in their position descriptions.

> 2.3.2.1 *Milestone: Fifty percent (50%) of contacted firms agree to include PtD skills in their position descriptions.*

"Business leaders' requirements for engineers and designers that are skilled in applying prevention through design principles can drive changes to the education of engineers and designers, both in the university as well as the continuing education setting. This will require the education of business leaders that Prevention through Design makes good business sense and will result in lower worker injury and illness rates, lower healthcare and worker compensation costs and improved worker satisfaction."

[Mann JA 2008]

2.3.3 Activity/Output Goal: By 2012, raise PtD awareness with the Society of Human Resource Management (SHRM) and Risk and Insurance Management Society (RIMS).

2.3.4 Activity/Output Goal: By 2011, develop a social marketing campaign demonstrating the value of integrating PtD principles into business management decisions by developing sector-specific social marketing messages about the value of preventing injuries, illnesses, fatalities, and exposures to the employees.

2.3.5 Activity/Output Goal: Develop a dissemination plan for the social marketing messages that reaches businesses in each industrial sector.

> 2.3.5.1 *Milestone: By 2011, a sector-specific social marketing campaign will be developed and disseminated. Effectiveness will be assessed through existing PtD feedback mechanisms.*

2.3.6 Activity/Output Goal: By 2011, develop a PtD message for the media through "Discoveries and Breakthroughs inside Science" (DBIS), focusing on the role of the engineer in reducing worker injuries, illnesses, and fatalities through design.

> 2.3.6.1 *Milestone: Five separate media publications carry the PtD message.*

2.3.7 Activity/Output Goal: Identify and influence business leaders on boards of directors or advisory boards of colleges of engineering by sharing information about PtD.

2.4 Intermediate Goal 4: Develop an educational compendium of successful PtD case studies and practices.

Performance Measure: By 2012, an electronic compendium of successful PtD practices will be available for integration into Engineering and H&S courses.

2.4.1 Activity/Output Goal: Collect successful PtD case studies and practices from the Sectors and Cross Sectors and evaluate the merit of the submitted practice.

2.4.2 Activity/Output Goal: Determine format(s) for the compendium.

2.4.3 Activity/Output Goal: Disseminate the compendium to target audiences.

2.5 Intermediate Goal 5: Include PtD concepts in certification exams.

Performance Measure: By 2014, Board of Certified Safety Professionals (BCSP), American Board of Industrial Hygiene (ABIH), and The Institute of Professional Environmental Practice (IPEP) all include PtD principles in their certification exams.

2.5.1 Activity/Output Goal: Work with the BCSP, ABIH, and IPEP to include PtD principles in certification exams.

2.5.2 Activity/Output Goal: Gain agreement from the BCSP and ABIH for delegates to receive certification maintenance points for attending educational programs on PtD.

3. Practice Goals

Strategic Goal: Stakeholders access, share, and apply successful PtD practices.

3.1 Intermediate Goal 1: Develop and disseminate a mechanism for identifying and sharing new and existing successful PtD practices.

Performance Measure: By 2011, a web-based application is developed and deployed that meets the needs of the NORA industrial sectors for obtaining information about successful PtD practices.

3.1.1 Activity/Output Goal: By 2010, create a system for web dissemination or expand or link to an existing Web-based system (e.g., Workplace Solutions database or, Center for Construction Research and Training solutions database).

3.1.2　Activity/Output Goal: Develop criteria to evaluate the merit of successful PtD practices and propose mechanism for collecting, evaluating, and uploading information, and identify the resources to manage the process.

> 3.1.2.1　Milestone: By 2011, disseminate criteria to evaluate the merit of successful PtD practices.

3.1.3　Activity/Output Goal: Develop templates for the NORA sectors to collect and share information. Develop search criteria.

> 3.1.3.1　Milestone: By 2011, disseminate templates for the NORA sectors to collect and share information.

> *"While there are many innovative ideas for encouraging use of PtD concepts, thought should be given to the best ways to communicate lessons learned, share best practices, and support diffusion of innovation. Ideas and concepts for PtD should be organized by industry, with best practices listed for each. Consideration should be given to the idea of a governing or organizing board to set priorities and format, which consists of representatives from each industry. This board will be responsible for creating and maintaining the database of best practices."*
>
> [Lin ML 2008]

3.1.4　Activity/Output Goal: Beta-test and deploy the application.

> 3.1.4.1　Milestone: By 2011, complete Beta test and deploy the application.

3.1.5　Activity/Output Goal: Investigate how promising NIOSH-developed control technologies (engineering design solutions) can be transferred from research into practice *(from the NIOSH/PtD NORA Research Project)*.

> 3.1.5.1　Milestone: From a list of six effective design solutions, select three for development of diffusion/dissemination strategies, including the development of business cases. Test and evaluate the success of the dissemination strategy. Timeline from 2010 through 2012.

3.2　Intermediate Goal 2: Develop and disseminate a mechanism for sharing design-related surveillance information.

Performance Measure: By 2012, mechanisms are in place that include design-related factors in incident investigations and provide the ability to track the incidence of these factors, by incident type and industrial sector, over time.

3.2.1 Activity/Output Goal: Using research results from Intermediate Goal 1.2, form a workgroup that proposes PtD surveillance needs to BLS.

3.2.2 Activity/Output Goal: Revise NIOSH Fatality Assessment and Control Evaluation (FACE) reports to include design-related factors.

3.2.3 Activity/Output Goal: Develop a mechanism for sharing industry notices of design-related factors that can cause or have caused serious injuries, illnesses, and fatalities.

3.2.4 Activity/Output Goal: Partner with FDA MedWatch (The Food and Drug Administration Safety Information and Adverse Event Reporting program) to include worker health and safety issues in its reports.

3.2.5 Activity/Output Goal: Include design-related factors in OSHA Fatal Facts.

3.2.6 Activity/Output Goal: Include information about PtD as it relates to tools and equipment in the US Consumer Product Safety Commission "Business" Web site.

3.2.7 Activity/Output Goal: Determine availability of design-related factors in the Consumer Product Safety Commission National Electronic Injury Surveillance System (NEISS). Add design-related factors, if absent.

3.2.8 Activity/Output Goal: Evaluate the use of blogs, YouTube, and other electronic media to share design-related surveillance information. Develop and implement a process to disseminate this information.

3.3 Intermediate Goal 3: Develop a system for identifying tools and equipment that includes design features to minimize risk of injury and illness. Expand from tools and equipment to facilities, processes, and organization of work.

Performance Measure: By 2013, a method to identify tools and equipment that include design factors to eliminate hazards and minimize risks will be developed and deployed. The method will include evaluation criteria that are accepted by tool and equipment trade associations as well as a label that is understood by consumers of these tools and equipment.

3.3.1 Activity/Output Goal: Evaluate developing the PtD equivalent of the "Energy Star" designation. Contact the EPA/DOE to learn more about how Energy Star was implemented.

3.3.2 Activity/Output Goal: Investigate potential barriers toward the concept of "safety" certification of tools.

3.3.3 Activity/Output Goal: Review existing criteria for manual and power tools in resources such as: *Workers and Their Tools: A Guide to the*

Ergonomic Design of Handtools and Small Presses, by Led Greenberg and Don B. Chaffin, and *Ergonomics of Power Tools*, by NIOSH and Rob Radwin.

3.3.4   Activity/Output Goal: Determine the scope and evaluation criteria.

3.3.5   Activity/Output Goal: Determine the appropriate agency/organization(s) to certify designs and manage/maintain the overall process.

3.3.6   Activity/Output Goal: Disseminate to all industrial sectors. Develop a public awareness campaign of the value of this designation.

3.4   *Intermediate Goal 4: Develop and disseminate methods to include health and safety specifications in the procurement of workplace tools and equipment.*

Performance Measure: By 2014, representative companies from each industrial sector have integrated health and safety specifications into the procurement of workplace tools and equipment and share their programs as "successful PtD practices."

3.4.1   Activity/Output Goal: Evaluate existing industry standards that include this requirement (e.g., ANSI/PMMI B155.1 Safety Standard for Packaging and Packaging-Related Converting Machinery).

3.4.2   Activity/Output Goal: Investigate existing programs, either at group purchasing organizations (GPOs) or industries, to serve as models for integrating health and safety specifications into purchasing decisions.

3.4.3   Activity/Output Goal: Determine opportunities and mechanisms, including materials evaluation tools, to improve group purchasing organizations to include more PtD information to assist procurement professionals in identifying equipment and tools that incorporate PtD features.

3.4.4   Activity/Output Goal: Evaluate the Department of Defense Design Criteria Standard for Human Engineering MIL-STD-1472 as a potential model for the integration of H&S design requirements into purchasing standards.

3.5   *Intermediate Goal 5: Demonstrate the value of worker involvement in health and safety design aspects of work areas, tools, and tasks.*

Performance Measure: By 2012, the value of employee involvement will be demonstrated through white paper publications, the publication of successful practices within industry, and incorporation in appropriate technical reports and standards.

3.5.1 Activity/Output Goal: Evaluate employee involvement in ergonomic tool design successes for possible diffusion to other H&S design applications.

> *"Worker input is critical in the design of work tasks and equipment."*
>
> [Johnson 2008]

3.5.2 Activity/Output Goal: Evaluate the potential for incorporating PtD considerations for worker involvement into "task-based risk assessment" models, e.g., ANSI B11.TR3.

3.5.3 Activity/Output Goal: Meet with representatives from UNITE HERE to evaluate the potential for transferring their success at worker involvement to other industries and industrial sectors.

3.5.4 Activity/Output Goal: Evaluate programs at companies that have successfully included workers in the health and safety design aspects of their work areas, tools, and tasks.

3.6 *Intermediate Goal 6: Demonstrate the value of PtD in achieving "lean" manufacturing.*

Performance Measure: By 20011, publish "safe and lean" project examples using existing NIOSH mechanisms.

3.6.1 Activity/Output Goal: Promote "safe and lean" concepts such as those outlined in ANSI B11 TR7 on Designing for Safety and Lean Manufacturing.

3.6.2 Activity/Output Goal: Collaborate with the Safe and Lean Network to develop templates and mechanisms for sharing successful "lean" H&S projects.

3.6.3 Activity/Output Goal: Develop a mechanism to obtain and promote Safety & Health Six Sigma, Design for Six Sigma, and Lean Manufacturing projects to demonstrate the goals of PtD. Include successful projects in proposed PtD communications mechanisms.

3.6.4 Activity/Output Goal: Introduce the concepts of PtD to the Six Sigma training and certification organizations.

3.7 *Intermediate Goal 7: For each business sector, define PtD and the steps to influence the development of a culture that promotes preventing occupational injuries, illnesses, fatalities, and exposures by designing out hazards and minimizing risks.*

Performance Measure: By 2014, PtD will be part of the organizational culture in OHSAS 18001-certified and OSHA VPP companies, as evidenced by their successful incorporation of PtD elements into their Health and Safety Management Systems.

3.7.1 Activity/Outcome Goal: Review PtD components of OHSAS 18001/2:1999, ANSI/AIHA Z10-2005 Occupational Health and Safety Management Systems (OHSMS), OSHA Voluntary Protection Programs (VPP), and OSHA Safety and Health Management System e-Tool. Recommend the inclusion of additional PtD program elements, as appropriate.

3.7.2 Activity/Outcome Goal: Review a representative sample of the commercially available health and safety management systems; include behavior-based safety systems, for the level of inclusion of PtD principles.

3.7.2.1 Milestone: By 2011, publish a white paper with the results and any recommendations.

3.7.3 Activity/Outcome Goal: By 2010, identify companies or organizations with well-developed PtD programs. Define the core elements, including performance metrics that have created a culture of PtD within these organizations.

3.7.3.1 Milestone: By 2012, publish results.

3.7.4 Activity/Outcome Goal: Develop a PtD award process, using Minerva Canada and the Applied Ergonomics Community ERGO Cup and other H&S design awards as models.

3.7.5 Activity/Output Goal: The level of adoption of PtD concepts among a subset of Fortune 500 companies in the United States through a survey conducted by ORC Worldwide will identify the level of leadership commitment to PtD among safety-conscious corporations and determine if implementing PtD concepts is associated with the companies' occupational injury and illness experience. The results of this survey will be used to develop an article that will be published in business journals *(from the NIOSH/PtD NORA Research Project)*.

3.8 Intermediate Goal 8: Incorporate PtD concepts into the business executive culture through a portfolio of PtD publications.

Performance Measure: By 2012, information about the business and social value of PtD will be available in publications read by business leaders.

3.8.1 Activity/Output Goal: Author three *papers* that are published in executive journals (e.g., *AG Executive, CEO Journal, Forbes Magazine, Fortune, Business Week, Harvard Business Review*)

3.9 Intermediate Goal 9: Develop a portfolio of PtD publications for workers and labor organizations.

Performance Measure: By 2012, information about the value of PtD for workers as well as their role in the PtD process will be available in publications or media read by workers.

> 3.9.1 Activity/Output Goal: Develop PtD articles, brochures, and/or topic pages for each industrial sector that are focused on the value of PtD to the worker as well as the worker's role in successful implementation of PtD at his/her workplace.

3.10 Intermediate Goal 10: Include preventive design aspects into the development of new technologies, work methods, and processes.

Performance Measure: By 2012, methods to design appropriate controls into nanotechnology processes will be developed and published.

> 3.10.1 Activity/Output Goal: Incorporate PtD principles in the development of guidelines for the safe handling of engineered nanoparticles.

> 3.10.2 Activity/Output Goal: Integrate safety and health into green elements of contractor specifications.

> 3.10.3 Activity/Output Goal: Develop index of "green" chemicals and exposure risks.

3.11 Intermediate Goal 11: Expand the scope of existing injury, illness, and fatality investigations and surveillance to include the identification of design-related factors and tracking of frequency and severity.

Performance Measure: By 2011, incident investigation methods will include the identification of design-related causal factors. By 2014, design-related injuries, illnesses, exposures, and fatalities will demonstrate a statistically significant reduction.

> 3.11.1 Activity/Output Goal: Develop a plan and implement recommendations from Intermediate Goal 1.2 and Activity/Output Goals 1.24 and 1.25.

3.12 Intermediate Goal 12: Develop a holistic, comprehensive, and practical framework to guide the project development process.

> 3.12.1 Activity/Output Goal: Develop transparent risk communication that includes OSH performance metrics.

> 3.12.2 Activity/Output Goal: Apply PtD principles to avoid risk transfer.

4. Policy Goals

Strategic Goal: Business leaders, labor, academics, government entities, and standard-developing and setting organizations endorse a culture that includes PtD principles in all designs affecting workers.

4.1   *Intermediate Goal 1: Develop and approve a broad and generic voluntary consensus standard on PtD that is aligned with international prevention through design activities and practices.*

Performance Measure: By 2012 consensus standard draft will include PtD principles as outlined by stakeholders and will provide a roadmap for industrial-sector or industry specific standards.

4.1.1   Activity/Output Goal: Support the development of an ANSI and/or ISO PtD standard.

4.2   *Intermediate Goal 2: Integrate PtD requirements into new and existing standards, codes, regulations and guidelines.*

Performance Measure: By 2012, a PtD consensus standard will be in progress; by 2011, at least five new and existing standards will include PtD concepts.

4.2.1   Activity/Output Goal: Understand ASSE, AIHA, and the NSC positions on PtD. Inform the boards of these and other relevant NGOs about the goals of the national initiative. Gain confirmation from their boards that efforts to include PtD into new and existing standards, codes, and guidelines will be considered.

4.2.2   Activity/Output Goal: Develop a list of existing health and safety or related standards to influence.

4.2.3   Activity/Output Goal: Develop an overview of PtD goals that can be considered for inclusion into standards. Provide these goals to NGOs and other organizations whose members serve on standards-development committees.

4.2.4   Activity/Output Goal: Provide access to PtD informational sessions and printed material to groups or individuals serving on standards-development committees.

4.2.5   Activity/Output Goal: Provide PtD information to the Standards Engineering Society (SES).

4.2.6   Activity/Output Goal: Develop model language for PtD goals that can serve as a tool for organizations or individuals that serve on standards-developing committees. Ensure model language is created in collaboration with ASSE, AIHA, NSC, and other NGOs as well as industry trade associations such as the Association of Equipment Manufacturers (AEM) and ORC WorldWide™. Include a review by product liability experts.

4.2.7   Activity/Output Goal: Develop an "A" Level standard for the ANSI B11 TR7 on Designing for Safety and Lean Manufacturing.

4.2.8 Activity/Output Goal: Review applicable Building Officials and Code Administrators (BOCA) requirements to determine the opportunities to include PtD concepts into these requirements.

4.2.9 Activity/Output Goal: Incorporate PtD principles into appropriate OSHA guidance topics.

4.2.10 Activity/Output Goal: Integrate PtD concepts and principles into management systems and consensus standards *(from the NIOSH/ PtD NORA Research Project)*.

> *4.2.10.1 Milestone: By 2011, compile summary reports/complete articles about the successful PtD incorporations and problems encountered.*

4.3 Intermediate Goal 3: The public and private sectors adopt PtD principles and methods in contracts.

Performance Measure: Government capital projects include all elements of PtD and are benchmark programs.

4.3.1 Activity/Output Goal: Include health and safety requirements in government tool and equipment specifications.

4.3.2 Activity/Output Goal: Include health and safety in the new and renovated facility design and construction processes to assure worker health and safety during construction, occupancy, and maintenance of these facilities.

4.3.4 Activity/Output Goal: Publish government building case studies as examples of effective, successful processes.

4.4 Intermediate Goal 4: Worker health and safety methods are included in sustainable design and construction practices.

Performance Measure: By 2014, "sustainable design" will include elements to eliminate hazards and minimize risk to workers.

4.4.1 Activity/Output Goal: Author a white paper on health and safety as a "sustainable" process. Publish paper to reach designers, engineers, and environmental professionals.

4.4.2 Activity/Output Goal: Work with US Green Building Council to include PtD principles in LEED° certification.

4.4.3 Activity/Output Goal: Integrate PtD principles into ISO 26000 on Corporate Responsibility.

4.4.4 Activity/Output Goal: Partner with the American Institute of Architects (AIA) to evaluate the potential for including elements to protect worker health and safety into the AIA Integrated Project Delivery Guidelines and Sustainability Resource Center.

4.4.5　Activity/Output Goal: Evaluate specific green design elements for potential H&S hazards/risks. Develop a plan to reduce potential hazards/risks.

4.4.6　Activity/Output Goal: Partner with the DOE Interagency Sustainability Working Group (ISWG) to include PtD principles on the ISWG website.

4.4.7　Activity/Output Goal: Evaluate the adoption of California Green Building Standards Code, CCR, Title 24, Part 11 by the California Building Standards Commission to determine opportunities to include PtD.

4.4.8　Activity/Output Goal: Develop a life-cycle assessment tool that includes occupational safety and health.

4.4.9　Activity/Output Goal: Develop, validate, and disseminate a LEED®-like safety and health rating system.

4.4.10　Activity/Output Goal: Investigate safer green products (cradle-to-cradle; life-cycle analysis).

5. Small Business Goals

Strategic Goal: Small businesses have access to PtD resources that are designed for or adapted to the small business environment.

5.1　Intermediate Goal 1: Develop a Social Marketing campaign demonstrating the value of integrating PtD principles into the small business design process.

Performance Measure: By 2011, small business will have access to sector-specific social marketing messages about the value of PtD.

5.1.1　Activity/Output Goal: Develop sector-specific social marketing messages about the value of preventing injuries, illnesses, fatalities, and exposures to the employees of small businesses.

5.1.2　Activity/Output Goal: Develop a dissemination plan for social marketing messages that reaches small businesses in each industrial sector.

5.2　Intermediate Goal 2: Develop case studies demonstrating savings to small business by implementing PtD.

Performance Measure: By 2011, business cases will be developed for applying PtD principles to representative H&S hazards and risks in the small business environment.

5.2.1　Activity/Output Goal: Establish and demonstrate the cost-avoidance links between worker and public safety for small business facilities that are open to the public, such as restaurants, retail sales, and services.

5.2.2 Activity/Output Goal: Demonstrate the improvements to small business in terms of quality, efficiency, and work organization that can be realized with the implementation of PtD.

5.3 Intermediate Goal 3: Promote alliances with economic development organizations to fund the implementation of effective PtD elements.

5.3.1 Activity/Output Goal: Investigate the potential for securing funding for PtD projects through small business loans, economic development organizations, local business alliances, and trade organizations. Provide the results of this information to small business using the dissemination methods resulting from activities in Intermediate Goal 5.5.

5.4 Intermediate Goal 4: Utilize the supply chain to influence the implementation of PtD.

Performance Measure: By 2013, a method to identify tools and equipment that include design factors to eliminate hazards and minimize risks will be developed and deployed. The method will include evaluation criteria that are accepted by tool and equipment trade associations as well as a label that is understood by consumers of these tools and equipment.

5.4.1 Activity/Output: Evaluate developing the PtD equivalent of the "Energy Star" designation. (See Activity/Output 3.3.1 through 3.3.6.) Develop a social marketing campaign to inform the general public—and therefore, the small business person—about the value of the designation in terms of preventing injuries, illnesses, and exposures as well as for enhancing business organization and efficiency.

5.4.2 Activity/Output: Evaluate opportunities to include health and safety design objectives into supply chain management by including provisions for the exchange of PtD information within the supply chain in relevant H&S management consensus standards.

5.5 Intermediate Goal 5: Develop and disseminate PtD information to small business.

Performance Measure: By 2012, small businesses will have access to low-cost PtD educational seminars and training material in 10% of large urban centers (>100,000 population). By 2014, small businesses will have access to low-cost PtD educational seminars and training materials in 50% of large urban centers. Small business persons in other areas will have access to information through Web-based dissemination methods.

5.5.1 Activity/Output: Develop educational program materials on PtD that can be used by workforce development programs, community colleges, and chambers of commerce.

5.5.2 Activity/Output Goal: Include PtD principles in the Workers' Compensation Pool Risk Management Guidelines.

5.5.3 Activity/Output Goal: Develop and disseminate Web-based PtD videos and informational pages that are specific to the needs of small business owners in each industrial sector.

5.6 *Intermediate Goal 6: Adapt successful PtD programs to the small business environment.*

Performance Measure: By 2011, Small Businesses will have web-based access to existing successful PtD solutions, adapted to the small business environment. By 2011 a process to receive, review, and disseminate new solutions, including business value calculations, will be established.

5.6.1 Activity/Output Goal: Review successful PtD practices for their transferability to small businesses.

5.6.2 Activity/Output Goal: Populate the Workplace Solutions database with simple, practical, transparent solutions for small businesses.

5.6.3 Activity/Output Goal: Identify designs that are affordable for smaller enterprises. Include those designs in the Workplace Solutions database.

5.7 *Intermediate Goal 7: Address the safety and health implications of purchasing previously owned or leased tools and equipment.*

Performance Measure: By 2013, a campaign to raise awareness about the potential hazards and risks associated with purchasing used tools and equipment will be developed and disseminated.

5.7.1 Activity/Output: Develop a plan for advising purchasers of resold equipment about design shortfalls that could affect the health and safety of affected workers.

5.8 *Intermediate Goal 8: Conduct surveillance and hazard research on green-job sites, with emphasis on small business.*

REFERENCES

Bealko SB, Kovalchik PG, Matetic RJ [2008]. Mining sector. J Safety Res *39*(2):187–189.

Behm M [2008]. Construction sector. J Safety Res *39*(2):175–178.

BLS [2008]. Injuries, illnesses, and fatalities. Industry injury and illness data. Washington, DC: U.S. Department of Labor, Bureau of Labor Statistics, Safety and Health Statistics Program. [www.bls.gov/iif/oshsum.htm].

Cervarich MB [2008] Prevention through partnerships. PtD in Motion, Issue 2.

Chapman LJ, Husberg B [2008]. Agriculture, forestry, and fishing sector. J Safety Res *39*(2):171–173.

Driscoll T, Harrison JE, Bradley C, Newson RS [2008]. The role of design issues in work-related fatal injury in Australia. J Safety Res *39*(2):209–214.

Fisher JM [2008]. Healthcare and social assistance sector. J Safety Res *39*(2):179–181.

Gambatese JA [2008]. Research issues in prevention through design. J Safety Res *39*(2):153–156.

Heidel DS [2008]. Manufacturing sector. J Safety Res *39*(2):183–186.

Howe J [2008}. Policy issues in prevention through design. J Safety Res *39*(2):161–163.

Johnson JV [2008]. Services Sector. J Safety Res *39*(2):191–194.

Kovalchik PG, Matetic RJ, Smith, AK, Bealko SB [2008]. Application of prevention through design for hearing loss in the mining industry. J Safety Res *39*(2):251–254.

Lin ML [2008]. Practice issues in prevention through design. J Safety Res *39*(2):157–159.

Lincoln JM, Lucas DL, McKibbin RW, Woodward CC, Bevan JE [2008]. Reducing commercial fishing deck hazards with engineering Solutions for winch design. J Safety Res *39*(2):231–235.

Madar SA [2008]. Transportation, warehousing, and utilities sector. J Safety Res *39*(2):195–197.

Mann JA [2008]. Education issues in prevention through design. J Safety Res *39*(2):165–170.

Mroszczyk JW [2008]. Wholesale and retail trade sector. J Safety Res 39(2):199–201.

NIOSH [2006]. Safe lifting and movement of nursing home residents. Cincinnati, OH: U.S. Department of Health and Human Services, Centers for Disease Control and Prevention, National Institute for Occupational Safety and Health, DHHS (NIOSH) Publication Number 2006–177.

Schulte PA [2005]. Characterizing the burden of occupational injury and disease. J Occup Environ Med 47:607–622.

Schulte PA, Rinehart R, Okun A, Geraci C, Heidel DS [2008]. National prevention through design (PtD) initiative. J Safety Res 39(2):115–121.

ADDITIONAL RESOURCES

AIHA (American Industrial Hygiene Association) [2008]. Demonstrating the business value of industrial hygiene. Methods and findings from the value of the industrial hygiene profession study. http://www.aiha.org/votp/report/VotP_Report.pdf.

American Society of Safety Engineers [1994]. ASSE position paper on designing for safety. Des Plaines, IL.

The Association for Manufacturing Technology [2000]. Risk assessment and reduction—a guide to estimate, evaluate and reduce risks associated with machine tools. McLean, VA, ANSI B11.TR3-2000.

Australian Government, The Australian Safety and Compensation Council [2008]. Safe design. http://www.ascc.gov.au/ascc/healthsafety/safedesign

Australian Government, The Australian Safety and Compensation Council [2006]. Guidance on the principles of safe design for work. Canberra, Australia.

Center for Construction Research and Training [2008]. http://www.cpwr.com

Design for Construction Safety [2009]. http://www.designforconstructionsafety.org

Christensen WC, Manuele FA [1999]. Safety through design. Itasca, IL: National Safety Council.

Manuele FA [2008]. Advanced safety management: Focusing on Z10 and serious injury prevention. Hoboken, NJ: John Wiley & Sons.

DOD (Department of Defense) [2000]. Standard practice for system safety, Mil-STD-882D. Washington, DC: Department of Defense. http://www.safetycenter.navy.mil/instructions/osh/milstd882d.pdfsearch='MILSTD882D'

UK Health and Safety Executive [2007]. Construction (Design and Management) Regulations 2007. http://www.hse.gov.uk/construction/cdm.htm

Packaging Machinery Manufacturers Institute [2006]. ANSI/PMMI B155.1-2006. Safety Requirements for Packaging Machinery and Packaging-Related Converting Machinery. Arlington, VA.

NOTES

NOTES

NOTES